Frederick Thomas Hodgson

A New System of Hand-Railing

Frederick Thomas Hodgson

A New System of Hand-Railing

ISBN/EAN: 9783744673990

Printed in Europe, USA, Canada, Australia, Japan

Cover: Foto ©berggeist007 / pixelio.de

More available books at **www.hansebooks.com**

OF

HAND-RAILING.

CUT SQUARE TO THE PLANK, WITHOUT THE AID OF FALLING MOULDS.

A NEW AND EASY METHOD OF FORMING HAND-RAILS.

Containing a Large Number of Illustrations of Hand-Rails,
with Full Instructions for Working Them.

BY

AN OLD STAIR-BUILDER.

REVISED AND CORRECTED EDITION

NEW YORK:
THE INDUSTRIAL PUBLICATION COMPANY.
1885.

PREFACE.

THE following method of getting out wreaths for hand-railing, while not new, is not generally known in this country, which is to be regretted. By this method much time and material are saved, and the wreaths formed with more accuracy than cutting them out by the older methods, as any wreath whatever may be cut out square from a plank of the same thickness as the diameter of a circle described round a section of the rail.

In the annexed cut, A B shows the thickness of the plank. But it is better for the learner to allow the plank ¼ inch thicker, as it leaves more for

squaring the rail. By this method the same bevel applies to *both ends of the wreath on the square joint,* for stairs with any number of winders in the well; this ensures the rail *rising equally all round the well,* and not quick in one part and flat in another.

All wreaths are cut out square from the plank with square joints, and so accurate are the bevels and joints that the rail may be cleaned up ready for polishing before it leaves the bench.

Seven complete examples are given, with ample illustrations and descriptive text; these examples being quite sufficient to enable the student to build a rail over any kind of a stair, no matter how many winders there may be, or how they are arranged.

New York, 1885.

INTRODUCTION.

BEFORE proceeding to describe the method of obtaining the lines necessary for the correct formation of Hand-rails, it is deemed advisable to present for the student's use a rudimentary treatise on what is known as CARPENTER'S GEOMETRY, as it is felt that a knowledge of this will materially aid the young beginner to a better understanding of the principles involved in the work.

The following treatise, which is taken from "Practical Carpentry," is simple and easily understood. Let me add, however, that, although a knowledge of geometry will greatly help the student to understand the method herein described, such knowledge is not absolutely necessary, as the Art of Hand-railing, as here presented, may be efficiently acquired without it.

TABLE OF CONTENTS.

PART I.

HAND-RAILING.

PART I.—GEOMETRY.

EFORE a knowledge of geometry can be acquired, it will be necessary to become acquainted with some of the terms and definitions used in the science of geometry, and to this end the following terms and explanations are given, though it must be understood that these are only a few of the terms used in the science, but they are sufficient for our purposes:

1. A *Point* has position but not magnitude. Practically, it is represented by the smallest visible mark or dot, but geometrically understood, it occupies no space. The extremities or ends of lines are points; and when two or more lines cross one another, the places that mark their intersections are also points.

2. A *Line* has length, without breadth or thickness, and, consequently. a true geometrical line cannot be exhibited; for however finely a line may be drawn, it will always occupy a certain extent of space.

3. A *Superficies* or *Surface* has length and breadth, but no thickness. For instance, a shadow gives a very good representation of a superficies: its length and breadth can be measured; but it has no depth or substance. The quantity of space contained in any plane surface is called its area.

4. A *Plane Superficies* is a flat surface, which will coincide with a straight line in every direction.

5. A *Curved Superficies* is an uneven surface, or such as will not coincide with a straight line in all directions. By the term surface is generally understood the outside of any body or object; as, for instance, the exterior of a brick or stone, the boundaries of which are represented by lines, either straight or curved, according to the form of the object. We must always bear in mind, however, that the lines thus bounding the figure occupy no part of the surface; hence the lines or points traced or marked on any body or surface, are merely symbols of the true geometrical lines or points.

6. A *Solid* is anything which has length, breadth and thickness; consequently, the term may be applied to any visible object containing substance; but, practically, it is understood to signify the solid contents or measurement contained within the different surfaces of which any body is formed.

7. *Lines* may be drawn in any direction, and are termed straight, curved, mixed, concave, or convex lines, according as they correspond to the following definitions.

8. A *Straight Line* is one every part of which A————————B lies in the same direction between its extremities, Fig. 1. and is, of course, the shortest distance between two points, as from A to B, Fig. 1.

9. A *Curved Line* is such that it does not lie in a straight direction between its extremities, but is continually changing by inflection. It may be either regular or irregular.

10. A *Mixed* or *Compound Line* is composed of straight and curved lines, connected in any form.

11. A *Concave* or *Convex Line* is such that it cannot be cut by a straight line in more than two points; the concave or hollow side is turned towards the straight line, while the convex or swelling side looks away from it. For instance, the inside of a basin is concave—the outside of a ball is convex.

12. *Parallel Straight Lines* have no inclination, but are everywhere at an equal distance from each other; consequently they can never meet, though produced or continued to infinity in either or both directions. Parallel lines may be either straight or curved,

provided they are equally distant from each other throughout their extension.

13. *Oblique* or *Converging Lines* are straight lines, which, if continued, being in the same plane, change their distance so as to meet or intersect each other.

14. A *Plane Figure, Scheme, or Diagram*, is the lineal representation of any object on a plane surface. If it is bounded by straight lines, it is called a rectilineal figure; and if by curved lines, a curvilineal figure.

15. An *Angle* is formed by the inclination of two lines meeting in a point: the lines thus forming the angle are called the sides; and the point where the lines meet is called the *vertex* or *angular point.*

When an angle is expressed by three letters, as A B C, Fig. 2, the middle letter B should always denote the angular point: where

Fig. 2

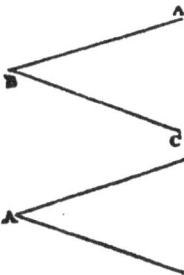
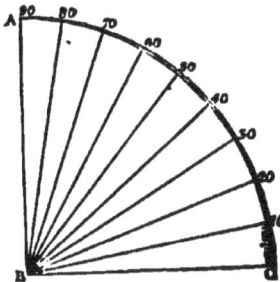

Fig. 3.   Fig. 4.   Fig. 5.

there is only one angle, it may be expressed more concisely by a letter placed at the angular point only, as the angle at A, Fig. 3.

16. The quantity of an angle is estimated by the arc of any circle contained between the two sides or lines forming the angle ; the junction of the two lines, or vertex of the angle, being the centre from which the arc is described. As the circumferences of all circles are proportional to their diameters, the arcs of similar sectors also bear the same proportion to their respective circumferences; and, consequently, are proportional to their diameters, and, of course, also to their radii or semi-diameters. Hence, the

proportion which the arc of any circle bears to the circumference
of that circle, determines the magnitude of the angle. From this
it is evident that the quantity or magnitude of angles does not de-
pend upon the length of the sides' or radii forming them, but
wholly upon the number of degrees contained in the arc cut from
the circumference of the circle by the opening of these lines. The
circumference of every circle is divided by mathematicians into
360 equal parts, called degrees; each degree being again subdi-
vided into 60 equal parts, called minutes, and each minute into 60
parts, called seconds. Hence, it follows that the arc of a quarter
circle or quadrant includes 90 degrees; that is, one-fourth part of
360 degrees. By dividing a quarter circle; that is, the portion of
the circumference of any circle contained between two radii forming a right angle, into 90 equal parts, or, as is shown in Fig. 4, into
nine equal parts of 10 degrees each, then drawing straight lines
from the centre through each point of division in the arc; the right
angle will be divided into nine equal angles, each containing 10
degrees. Thus, suppose B C the horizontal line, and A B the perpendicular ascending from it, any line drawn from B—the centre
from which the arc is described—to any point in its circumference,
determines the degree of inclination or angle formed between it
and the horizontal line B C. Thus, a line from the centre B to the
tenth degree, separates an angle of 10 degrees, and so on. In this
manner the various slopes or inclinations of angles are defined.

17. A *Right Angle* is produced by one straight line standing
upon another, so as to make the adjacent angles equal. This is
what workmen call "square," and is the most useful figure they
employ.

18. An *Acute Angle* is less than a right angle, or less than 90
degrees.

19. An *Obtuse Angle* is greater than a right angle or square, or
more than 90 degrees.

The number of degrees by which an angle is less than 90 degrees is called the complement of the angle. Also, the difference
between an obtuse angle and a semicircle, or 180 degrees, is called
the supplement of that angle.

20. *Plane Figures* are bounded by straight lines, and are named according to the number of sides which they contain. Thus, the space included within three straight lines, forming three angles, is called a trilateral figure or triangle.

21. A *Right-Angled Triangle* has one right angle: the sides forming the right angle are called the base and perpendicular; and the side opposite the right angle is named the hypothenuse. An equilateral triangle has all its sides of equal length. An isosceles triangle has only two sides equal; a scalene triangle has all its sides unequal. An acute-angled triangle has all its angles acute, and an obtused-angled triangle has one of its angles only obtuse.

The triangle is one of the most useful geometrical figures for the mechanic in taking dimensions; for since all figures that are bounded by straight lines are capable of being divided into triangles, and as the form of a triangle cannot be altered without changing the length of some one of its sides, it follows that the true form of any figure can be preserved if the length of the sides of the different triangles into which it is divided is known; and the area of any triangle can easily be ascertained by the same rule, as will be shown further on.

Quadrilateral Figures are literally four-sided figures. They are also called quadrangles, because they have four angles.

22. A *Parallelogram* is a figure whose opposite sides are parallel, as A B C D, Fig. 5.

23. A *Rectangle* is a parallelogram having four right angles, as A B C D, in Fig. 5.

24. A *Square* is an equilateral rectangle, having all its sides equal, like Fig. 5.

25. An *Oblong* is a rectangle whose adjacent sides are unequal, as the parallelogram shown at Fig. 10.

26. A *Rhombus* is an oblique-angled figure, or parallelogram having four equal sides, whose opposite angles only are equal, as c, Fig. 6.

27. A *Rhomboid* is an oblique-angled parallelogram, of which the adjoining sides are unequal, as D, Fig. 7.

.28. A *Trapezium* is an irregular quadrilateral figure, having no two sides parallel, as E, Fig. 8.

29. A *Trapezoid* is a quadrilateral figure, which has two of its opposite sides parallel, and the remaining two neither parallel nor equal to one another, as F, Fig. 9.

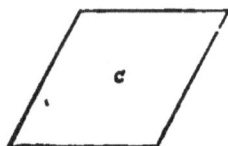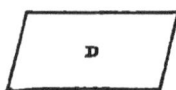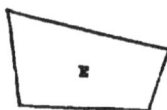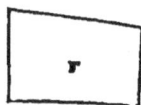

Fig. 6.   Fig. 7.   Fig. 8.   Fig. 9.

30. A *Diagonal* is a straight line drawn between two opposite angular points of a quadrilateral figure, or between any two angular points of ·a polygon. Should the figure be a parallelogram, the diagonal will divide it into two equal triangles, the opposite sides and angles of which will be equal to one another. Let A B C D, Fig. 10, be a parallelogram; join A C, then A C is a diagonal, and the triangles A D C, A B C, into which it divides the parallelogram, are equal.

31. A plane figure, bounded by more than four straight lines, is called a *Polygon*. A regular polygon has all its sides equal, and consequently its angles are also equal, as K, L, M, and N, Figs.

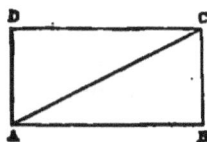

Fig. 10.   Fig. 11.   Fig. 12.   Fig. 13.

12–15. An irregular polygon has its sides and angles unequal, as H, Fig. 11. Polygons are named according to the number of their sides or angles, as follows :—

32. A *Pentagon* is a polygon of five sides, as H or K, Figs. 11, 12.

33. A *Hexagon* is a polygon of six sides, as L, Fig. 13.

34. A *Heptagon* has seven sides, as M, Fig. 14.

35. An *Octagon* has eight sides, as N, Fig. 15.

An *Enneagon* has nine, a *Decagon* ten, an *Undecagon* eleven, and a *Dodecagon* twelve sides. Figures having more than twelve sides are generally designated *Polygons*, or many-angled figures.

36. A *Circle* is a plane figure bounded by one uniformly curve 1 line, *b c d* (Fig. 16), called the circumference, every part of which is equally distant from a point within it, called the centre, as *a*.

37. The *Radius* of a circle is a straight line drawn from the centre to the circumference; hence, all the radii (plural for radius) of the same circle are equal, as *b a, c a, e a, f a*, in Fig. 16.

38. The *Diameter* of a circle is a straight line drawn through the centre, and terminated on each side by the circumference; conse-

 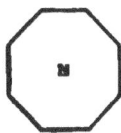 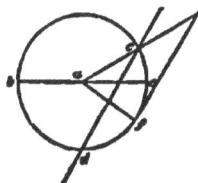 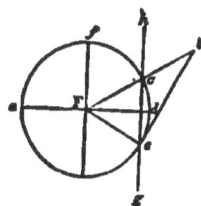

Fig. 14.   Fig. 15.   Fig. 16.   Fig. 17.

quently the diameter is exactly twice the length of the radius; and hence the radius is sometimes called the semi-diameter. (See *b a c*, Fig. 16.)

49. The *Chord* or *Subtens* of an arc is any straight line drawn from one point in the circumference of a circle to another, joining the extremities of the arc, and dividing the circle either into two equal, or two unequal parts. If into equal parts, the chord is also the diameter, and the space included between the arc and the diameter, on either side of it, is called a semicircle, as *b a e* in Fig. 16. If the parts cut off by the chord are unequal, each of them is called a *segment* of the circle. The same chord is therefore common to two arcs and two segments; but, unless when stated otherwise, it is always understood that the lesser arc or segment is spoken of, as in Fig. 16, the chord *c d* is the chord of the arc *c e d*.

If a straight line be drawn from the centre of a circle to meet the chord of an arc perpendicularly, as *a f*, in Fig. 16, it will divide the chord into two equal parts, and if the straight line be produced

to meet the arc, it will also divide it into two equal parts, as
c f, f d.

Each half of the chord is called the sine of the half-arc to which
it is opposite; and the line drawn from the centre to meet the
chord perpendicularly, is called the co-sine of the half-arc. Consequently, the radius, the sine, and co-sine of an arc form a right
angle.

40. Any line which cuts the circumference in two points, or a
chord lengthened out so as to extend beyond the boundaries of
the circle, such as *g h* in Fig. 17, is sometimes called a *Secant*.
But, in trigonometry, the secant is a line drawn from the centre
through one extremity of the arc, so as to meet the tangent which is
drawn from the other extremity at right angles to the radius.
Thus, *f c b* is the secant of the arc *c e*, or the angle *c f e*, in Fig. 17.

41. A *Tangent* is any straight line which touches the circumference of a circle in one point. which is called the point of contact,
as in the tangent line *e b*, Fig. 17.

42. A *Sector* is the space included between any two radii, and
that portion of the circumference comprised between them : *c e f'*
is a sector of the circle *a f c e*, Fig. 17.

43. A *Quadrant*, or quarter of a circle, is a sector bounded by
two radii, forming a right angle at the centre, and having one-
fourth part of the circumference for its arc, as *f f d*, Fig. 7.

44. An *Arc*, or *Arch*, is any portion of the circumference of a
circle, as *c d e*, Fig. 17.

It may not be improper to remark here that the terms circle and
circumference are frequently misapplied. Thus we say, describe a
circle from a given point, etc., instead of saying describe the circumference of a circle—the circumference being the curved line
thus described, everywhere equally distant from a point within it,
called the centre; whereas the circle is properly the superficial
space included within that circumference.

45. *Concentric Circles* are circles within circles, described from
the same centre; consequently, their circumferences are parallel to
one another, as Fig. 18.

46. *Eccentric Circles* are those which are not described from the

same centre; any point which is not the centre is also eccentric in reference to the circumference of that circle. Eccentric circles may also be tangent circles; that is, such as come in contact in one point only, as Fig. 19.

47. *Altitude.* The height of a triangle or other figure is called its *altitude*. To measure the altitude, let fall a straight line from the vertex, or highest point in the figure, perpendicular to the base or opposite side; or to the base continued, as at B D, Fig. 20, should the form of the figure require its extension. Thus C D is the altitude of the triangle A B C.

We have now described all the figures we shall require for the purpose of thoroughly understanding all that will follow in this book; but we would like to say right here that the student who has time should not stop at this point in the study of geometry, for the time spent in obtaining a thorough knowledge of this useful

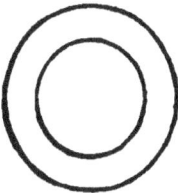

Fig. 18.

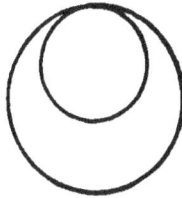

Fig. 19.

Fig. 20.

science will bring in better returns in enjoyment and money, than if expended for any other purpose.

We will now proceed to explain how the figures we have described can be constructed. There are several ways of constructing nearly every figure we produce, but we have chosen those methods that seemed to us the best, and to save space have given as few examples as possible consistent with efficiency.

PROBLEM I.—*Through a given point* C (Fig. 18 *a*), *to draw a straight line parallel to a given straight line* A B.

In A B (Fig. 18 *a*) take any point *d*, and from *d* as a centre with the radius *d* C, describe an arc C *e*, cutting A B in *e*, and from C as a

centre, with the same radius, describe the arc *d* D, make *d* D equal to
C *e*, join C D, and it will be parallel to A B.

PROBLEM II.—*To make an angle equal to a given rectilineal angle.*

From a given point E (Fig. 19 *a*), upon the straight line E F, to
make an angle equal to the given angle A B C. From the angular
point B, with any radius, describe the arc *e f*, cutting B C and B A in
the points *e* and *f*. From the point E on E F with the same radius,

Fig. 18 *a*.   Fig. 19 *a*.

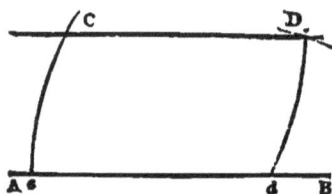

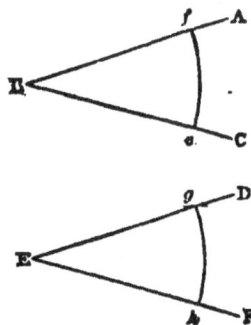

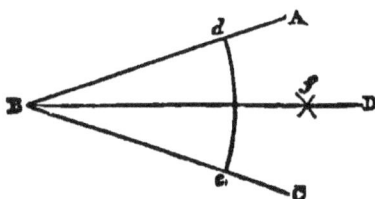

Fig. 20 *a*.

describe the arc *h g*, and make it equal to the arc *e f*; then from E,
through *g*, draw the line E D: the angle D E F will be equal to the
angle A B C.

PROBLEM III.—*To bisect a given angle.*

Let A B C (Fig. 20 *a*) be the given angle. From the angular point
B, with any radius, describe an arc cutting B A and B C in the
points *d* and *e*; also, from the points *d* and *e* as centres, with any
radius greater than half the distance between them, describe arcs
cutting each other in *f*; through the points of intersection *f*, draw
B *f* D: the angle A B C is bisected by the straight line B D; that is,
it is divided into two equal angles, A B D and C B D.

PROBLEM IV.—*To trisect or divide a right angle into three equal
angles.*

Let A B C (Fig. 21) be the given right angle. From the angular point B, with any radius, describe an arc cutting B A and B C in the points *d* and *g*; from the points *d* and *g*, with the radius B *d* or B *g*, describe the arcs cutting the arc *d g* in *e* and *f*; join B *e* and B *f*: these lines will trisect the angle A B C, or divide it into three equal angles.

The trisection of an angle can be effected by means of elementary geometry only in a very few cases; such, for instance, as those where the arc which measures the proposed angle is a whole circle, or a half, a fourth, or a fifth part of the circumference. Any angle of a pentagon is trisected by diagonals, drawn to its opposite angles.

PROBLEM V.—*From a given point* C, *in a given straight line* A B, *to erect a perpendicular.*

From the point C (Fig. 22), with any radius less than C A or C B, describe arcs cutting the given line A B in *d* and *e*; from these

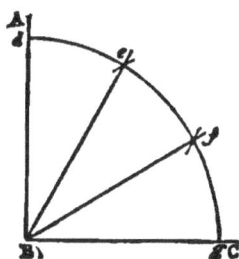
Fig. 21.

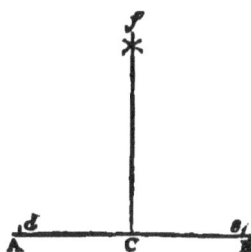
Fig. 22.

points as centres, with a radius greater than C *d* or C *e*, describe arcs intersecting each other in *f*: join C *f*, and this line will be the perpendicular required.

Another Method.—To draw a right angle or erect a perpendicular by means of any scale of equal parts, or standard measure of inches, feet, yards, etc., by setting off distances in proportion to the numbers 3, 4 and 5, or 6, 8 and 10, or any numbers whose squares correspond to the sides and hypothenuse of a right-angled triangle.

From any scale of equal parts, as that represented by the line D (Fig. 23), which contains 5, set off from B, on the line A B, the distance B *e*, equal to 3 of these parts; then from B, with a radius equal to 4 of the same parts, describe the arc *a b*; also from *e* as a

centre, with a radius equal to 5 parts, describe another arc intersecting the former in c; lastly join в c; the line в c will be perpendicular to a в.

This mode of drawing right angles is more troublesome upon
paper than the method previously given ; but in laying out grounds
or foundations of buildings it is often very useful, since only with a
ten-foot pole, tape line, or chain, perpendiculars may be set out
very accurately. The method is demonstrated thus :—The square
of the hypothenuse, or longest side of a right-angled triangle, being
equal to the sum of the squares of the other two sides, the same
property must always be inherent in any three numbers, of which

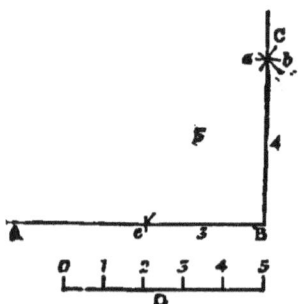

Fig. 23.

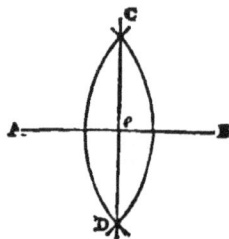

Fig. 24.

the squares of the two lesser numbers, added together, are equal to
the square of the greater. For example, take the numbers 3, 4,
and 5 ; the square of 3 is 9, and the square of 4 is 16 ; 16 and 9,
added together make 25, which is 5 times 5, or the square of the
greater number. Although these numbers, or any multiple of
them, such as 6, 8, 10, or 12, 16, 20, etc., are the most simple, and
most easily retained in the memory, yet there are other numbers,
very different in proportion, which can be made to serve the same
purpose. Let n denote any number; then $n^2 + 1$, $n^2 - 1$, and $2n$,
will represent the hypothenuse, base, and perpendicular of a right-
angled triangle.   Suppose $n = 6$, then $n^2 + 1 = 37$, $n^2 - 1 = 35$,
and $2n = 12$: hence, 37, 35, and 12 are the sides of a right-angled
triangle. A knowledge of this problem will often prove of the
greatest service to the workman.

PROBLEM VI.—*To bisect a given straight line.*

Let A B (Fig. 24) be the given straight line. From the extreme points A and B as centres, with any equal radii greater than half the length of A B, describe arcs cutting each other in C and D : a straight line drawn through the points of intersection C and D, will bisect the line A B in *e*.

PROBLEM VII.—*To divide a given line into any number of equal parts.*

Let A B (Fig. 25) be the given line to be divided into five equal parts. From the point A draw the straight line A C, forming any angle with A B. On the line A C, with any convenient opening of the compasses, set off five equal parts towards C; join the extreme

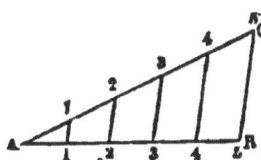 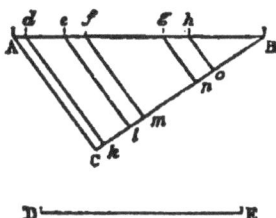 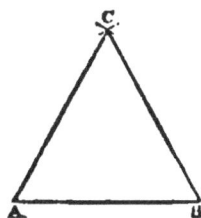

Fig. 25.   Fig. 26.   Fig. 27

points C B; through the remaining points 1, 2, 3, and 4, draw lines parallel to B C, cutting A B in the corresponding points, 1, 2, 3, and 4 : A B will be divided into five equal parts, as required.

There are several other methods by which lines may be divided into equal parts; they are not necessary, however, for our purpose, so we will content ourselves with showing how this problem may be used for changing the scales of drawings whenever such change is desired. Let A B (Fig. 26) represent the length of one scale or drawing, divided into the given parts A *d*, *d e*, *e f*, *f g*, *g h*, and *h* B; and D E the length of another scale or drawing required to be divided into similar parts. From the point B draw a line B C = D E, and forming any angle with A B; join A C, and through the points *d*, *e*, *f*, *g*, and *h*, draw *d k*, *e l*, *f m*, *g n*, *h o*, parallel to A C; and the parts C *k*, *k l*, *l m*, etc., will be to each other, or to the whole line B C, as the lines A *d*, *d e*, *e f*, etc., are to each other, or to the given line or scale A B. By this method, as will be evident from the figure, similar divisions can be obtained in lines of any given length.

PROBLEM VIII.—*To describe an equilateral triangle upon a given straight line.*

Let A B (Fig. 27) be the given straight line. From the points A and B, with a radius equal to A B, describe arcs intersecting each other in the point C. Join C A and C B, and A B C will be the equilateral triangle required.

PROBLEM IX.—*To construct a triangle whose sides shall be equal to three given lines,* F, E, D.

Draw A B (Fig. 28) equal to the given line F. From A as a centre, with a radius equal to the line E, describe an arc; then

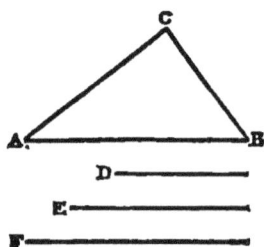

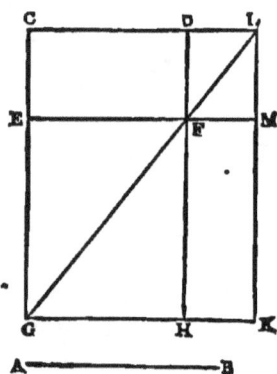

Fig. 28.   Fig. 29.

from B as a centre, with a radius equal to the line D, describe another arc intersecting the former in C; join C A and C B, and A B C will be the triangle required.

PROBLEM X.—*To describe a rectangle or parallelogram having one of its sides equal to a given line, and its area equal to that of a given rectangle.*

Let A B (Fig. 29) be the given line, and C D E F the given rectangle. Produce C E to G, making E G equal to A B; from G draw G K parallel to E F, and meeting D F produced in H. Draw the diagonal G F, extending it to meet C D produced in L; also draw L K parallel to D H, and produce E F till it meet L K in M; then F M K H is the rectangle required.

Equal and similar rhomboids or parallelograms of any dimensions may be drawn after the same manner, seeing the complements of the parallelograms which are described on or about the diagonal of any parallelogram, are always equal to each other; while the parallelograms themselves are always similar to each other, and to the original parallelogram about the diagonal of which they are constructed. Thus, in the parallelogram C G K L the complements C E F D and F M K H are always equal, while the parallelograms F F H G and D F M L about the diagonal G L, are always similar to each other, and to the whole parallelogram C G K L.

PROBLEM XI.—*To describe a square equal to two given squares.*

Let A and B (Fig. 30) be the given squares. Place them so that a side of each may form the right angle D C E; join D E, and upon this hypothenuse describe the square D E G F, and it will be equal to the sum of the squares A and B, which are constructed upon the legs of the right-angled triangle D C E. In the same manner, any other rectilineal figure, or even circle, may be found equal to the sum of other two similar figures or circles. Suppose the lines C D and C E to be the diameters of two circles, then D E will be the diameter of a third, equal in area to the other two circles. Or suppose C D and C E to be the like sides of any two similar figures, then D E will be the corresponding side of another similar figure equal to both the former.

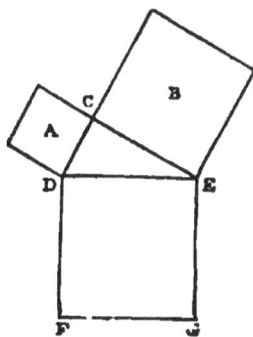
Fig. 30.

PROBLEM XII.—*To describe a square equal to any number of given squares.*

Let it be required to construct a square equal to the three given squares A, B, and C (Fig. 31). Take the line D E, equal to the side of the square C. From the extremity D erect D F perpendicular to D E, and equal to the side of the square B: join E F; then a square described upon this line will be equal to the sum of the two given squares C and B. Again, upon the straight line E F erect the per-

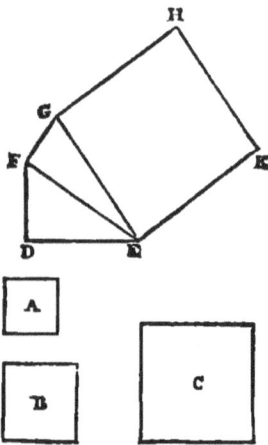

Fig. 31.

pendicular F G, equal to the side of the third given square A ; and join G E, which will be the side of the square G E H K, equal in area to A, B, and C. Proceed in the same way for any number of given squares.

PROBLEM XIII.—*Upon a given straight line to describe a regular polygon.*

To produce a regular pentagon draw A B to C (Fig. 32), so that B C may be equal to A B ; from B as a centre, with the radius B A or B C describe the semicircle A D C; divide the semi-circumference A D C into as many equal parts as there are parts in the required polygon, which, in the present case, will be five ; through the second division from C draw the straight line B D, which will form another side of the figure. Bisect A B at *e*, and B D at *f*, and draw *e* G and *f* G perpendicular to A B and B D; then G, the point of intersection, is the centre of a circle, of which A B and D are points in the circumference. From G, with a radius equal to its distance from any of these points, describe the circumference A B D H K; then producing the dotted lines from the centre B, through the remaining divisions in the semicircle

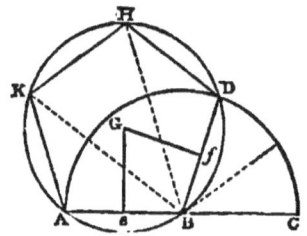

Fig. 32.

A D C, so as to meet the circumference of which G is the centre, in H and K, these points will divide the circle A B D H K into the number of parts required, each part being equal to the given side of the pentagon.

From the preceding example it is evident that polygons of any number of sides may be constructed upon the same principles, because the circumferences of all circles, when divided into the same number of equal parts, produce equal angles; and, consequently, by dividing the semi-circumference of any given circle into the number of parts required, two of these parts will form an angle which will be subtended by its corresponding part of the whole circumference. And as all regular polygons can be inscribed in a circle, it must necessarily follow, that if a circle be described through three given angles of that polygon, it will contain the number of sides or angles required.

The above is a general rule, by which all regular polygons may be described upon a given straight line; but there are other methods by which many of them may be more expeditiously constructed, as shown in the following examples:—

PROBLEM XIV.—*Upon a given straight line to describe a regular pentagon.*

Let A B (Fig. 33) be the given straight line; from its extremity B erect B c perpendicular to A B, and equal to its half. Join A c, and produce it till c d be equal to B c, or half the given line A B. From A and B as centres, with a radius equal to B d, describe arcs intersecting each other in e, which will be the centre of the circumscribing circle A B F G H. The side A B applied successively to this circumference, will give the angular points of the pentagon; and these being connected by straight lines will complete the figure.

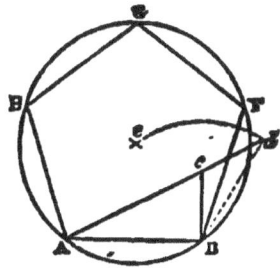

Fig. 33.

PROBLEM XV.—*Upon a given straight line to describe a regular hexagon.*

Let A B (Fig. 34) be the given straight line. From the extremities A and B as centres, with the radius A B describe arcs cutting each other in g. Again from g, the point of intersection, with the same radius, describe the circle A B c, which will contain the given side A B six times when applied to its circumference, and will be the hexagon required.

PROBLEM XVI.—*To describe a regular octagon upon a given straight line.*

Let A B (Fig. 35) be the given line. From the extremities A and B erect the perpendiculars A E and B F; extend the given line both ways to *k* and *l*, forming external right angles with the lines A E and B F. Bisect these external right angles, making each of the bisecting lines A H and B C equal to the given line A B. Draw H G and C D parallel to A E or B F, and each equal in length to A B. From G draw G E parallel to B C, and intersecting A E in E, and from D draw D F parallel to A H, intersecting B F in F. Join E F, and A B C D F E G H is the octagon required. Or from D

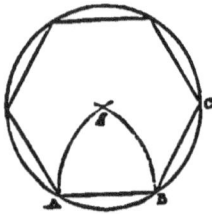 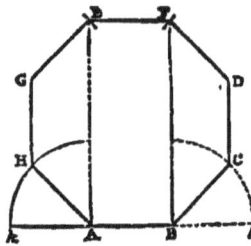 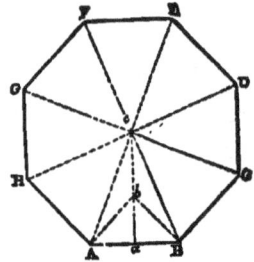

Fig 34.   Fig. 35.   Fig. 36.

and G as centres, with the given line A B as radius, describe arcs cutting the perpendiculars A E and B F in E and F, and join G E, E F, F D, to complete the octagon.

Otherwise, thus.—Let A B (Fig. 36) be the given straight line on which the octagon is to be described. Bisect it in *a*, and draw the perpendicular *a b* equal to A *a* or B *a*. Join A *b*, and produce *a b* to *c*, making *b c* equal to A *b*; join also A *c* and B *c*, extending them so as to make *c* E and *c* F each equal to A *c* or B *c*. Through *c* draw C *c* G at right angles to A E. Again, through the same point *c*, draw D H at right angles to B F, making each of the lines *c* C, *c* D, *c* G, and *c* H equal to A *c* or *c* B, and consequently equal to one another. Lastly, join B C, C D, D E, E F, F G, G H, H A; A B C D E F G H will be a regular octagon described upon A B, as required.

PROBLEM XVII.—*In a given square to inscribe a given octagon.*

Let A B C D (Fig. 37) be the given square. Draw the diagonals A C and B D, intersecting each other in *e*; then from the angular points A B C and D as centres, with a radius equal to half the diagonal, viz., A *e* or C *e*, describe arcs cutting the sides of the square in the points *f, g, h, k, l, m, n, o,* and the straight lines *o f, g h, k l,* and *m n,* joining these points will complete the octagon, and be inscribed in the square A B C D, as required.

PROBLEM XVIII.—*To find the area of any regular polygon.*

Let the given figure be a hexagon; it is required to find its area. Bisect any two adjacent angles, as those at A and B (Fig. 38), by the straight lines A C and B C, intersecting in C, which will be the

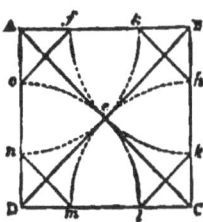
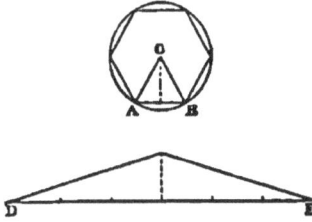
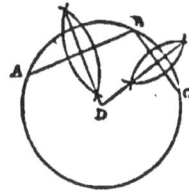

Fig. 37.   Fig. 38.   Fig. 39.

centre of the polygon. Mark the altitude of this elementary triangle by a dotted line drawn from C perpendicular to the base A B; then multiply together the base and altitude thus found, and this product by the number of sides : half gives the area of the whole figure.

Or otherwise, thus.—Draw the straight line D E, equal to six times, *i. e.,* as many times A B, the base of the elementary triangle, as there are sides in the given polygon. Upon D E describe an isosceles triangle, having the same altitude as A B C, the elementary triangle of the given polygon; the triangle thus constructed is equal in area to the given hexagon; consequently, by multiplying the base and altitude of this triangle together, half the product will be the area required. The rule may be expressed in other words, as follows :—The area of a regular polygon is equal to its perimeter, multiplied by half the radius of its inscribed circle, to which the sides of the polygon are tangents.

PROBLEM XIX.— *To describe the circumference of a circle through three given points.*

Let A, B, and C (Fig. 39) be the given points not in a straight line. Join A B and B C; bisect each of the straight lines A B and B C by perpendiculars meeting in D; then A, B and C are all equi-

distant from D; therefore a circle described from D, with the radius
D A, D B, or D C, will pass through all the three points as required.

PROBLEM XX.—*To divide a given circle into any number of equal
or proportional parts by concentric divisions.*

Let A B C (Fig. 40) be the given circle, to be divided into five
equal parts. Draw the radius A D, and divide it into the same

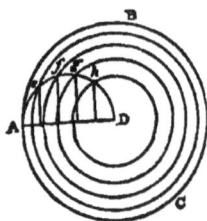 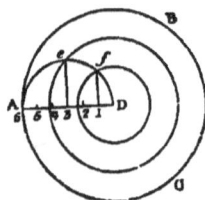 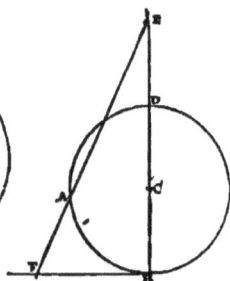

Fig. 40.   Fig. 41.   Fig. 42.

number of parts as those required in the circle; and upon the
radius thus divided, describe a semicircle: then from each point of
division on A D, erect perpendiculars to meet the semi-circumference in *e, f, g,* and *h.* From D, the centre of the given circle,
with radii extending to each of the different points of intersection
on the semicircle, describe successive circles, and they will divide
the given circle into five parts of equal area as required; the centre
part being also a circle, while the other four will be in the form of
rings.

PROBLEM XXI.—*To divide a circle into three concentric parts,
bearing to each other the proportion of one, two, three, from the centre.*

Draw the radius A D (Fig. 41), and divide it into six equal parts.
Upon the radius thus divided, describe a semicircle: from the first
and third points of division, draw perpendiculars to meet the semi-
circumference in *e* and *f.* From D, the centre of the given circle,
with radii extending to *e* and *f,* describe circles which will divide
the given circle into three parts, bearing to each other the same
proportion as the divisions on A D, which are as 1, 2 and 3. In
like manner circles may be divided in any given ratio by concentric
divisions.

PROBLEM XXII.—*To draw a straight line equal to any given arc of a circle.*

Let A B (Fig. 42) be the given arc. Find c the centre of the arc, and complete the circle A D B. Draw the diameter B D, and produce it to E, until D E be equal to C D. Join A E, and extend it so as to meet a tangent drawn from B in the point F; then B F will be nearly equal to the arc A B.

The following method of finding the length of an arc is equally simple and practical, and not less accurate than the one just given.

Let A B (Fig. 43) be the given arc. Find the centre c, and join A B, B C, and C A. Bisect the arc A B in D, and join also C D; then through the point D draw the straight line E D F, at right

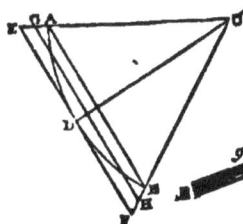

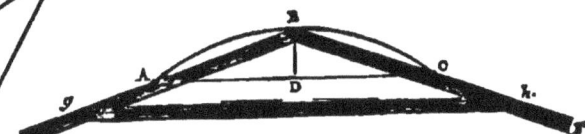

Fig. 43.   Fig. 44.

angles to C D, and meeting C A and C B produced in E and F. Again, bisect the lines A E and B F in the points G and H. A straight line G H, joining these points, will be a very near approach to the length of the arc A B.

Seeing that in very small arcs the ratio of the chord to the double tangent or, which is the same thing, that of a side of the inscribed to a side of the circumscribing polygon, approaches to a ratio of equality, an arc may be taken so small, that its length shall differ from either of these sides by less than any assignable quantity; therefore, the arithmetical mean between the two must differ from the length of the arc itself by a quantity less than any that can be assigned. Consequently the smaller the given arc, the more nearly will the line found by the last method approximate to the exact length of the arc. If the given arc is above 60 degrees, or two-thirds of a quadrant, it ought to be bisected, and the length of the semi-arc thus found being double, will give the length of the whole arc.

These problems are very useful in obtaining the lengths of veneers or other materials required for bending round soffits of door and window-heads.

PROBLEM XXIII.—*To describe the segment of a circle by means of two laths, the chord and versed sine being given.*

Take two rods, E B, B F (Fig. 44), each of which must be at least equal in length to the chord of the proposed segment A C: join them together at B, and expand them, so that their edges shall pass through the extremities of the chord, and the angle where they join shall be on the extremity B of the versed sine D B, or height of the segment. Fix the rods in that position by the cross piece *g h*, then by guiding the edges against pins in the extremities of the chord line A C, the curve A B C will be described by the point B.

PROBLEM XXIV.—*Having the chord and versed sine of the segment of a circle of large radius given, to find any number of points in the curve by means of intersecting lines.*

Let A C be the chord and D B the versed sine.

Through B (Fig. 45) draw E F indefinitely and parallel to A C; join A B, and draw A E at right angles to A B. Draw also A G at

Fig. 45.

right angles to A C, or divide A D and E B into the same number of equal parts, and number the divisions from A and E respectively, and join the corresponding numbers by the lines 1 1, 2 2, 3 3. Divide also A G into the same number of equal parts as A D or E B, numbering the divisions from A upwards, 1, 2, 3, etc.; and from the points , 2 and 3, draw lines to B ; and the points of intersection of these, with the other lines at *h, k, l*, will be points in the curve required. Same with B C.

Another Method.—Let A C (Fig. 46) be the chord and D B the versed sine. Join A B, B C, and through B draw E F parallel to A C.

From the centre B, with the radius B A or B C, describe the arcs
A E, C F. and divide them into any number of equal parts, as 1, 2,
3: from the divisions 1, 2, 3, draw radii to the centre B, and divide
each radius into the same number of equal parts as the arcs A E

Fig. 46.

and C F; and the points *g, h, l, m, n, o,* thus obtained, are points in
the required curve.

These methods, though not absolutely correct, are sufficiently
accurate when the segment is less than the quadrant of a circle.

PROBLEM XXV.—*To draw an ellipse with the trammel.*

The trammel is an instrument consisting of two principal parts,
the fixed part in the form of a cross E F G H (Fig. 47), and the
moveable piece or tracer *k l m.* The fixed piece is made of two
rectangular bars or pieces of wood, of equal thickness, joined together so as to be in the same plane. On one side of the frame
so formed, a groove is made, forming a right-angled cross. In the
groove two studs, *k* and *l,* are fitted
to slide freely, and carry attached to
them the tracer *k l m.* The tracer
is generally made to slide through a
socket fixed to each stud, and provided with a screw or wedge, by
which the distance apart of the studs
may be regulated. The tracer has
another slider *m,* also adjustable,

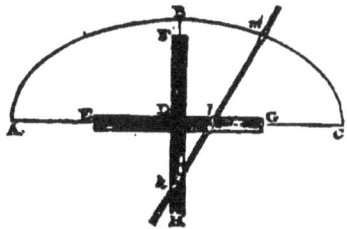

Fig. 47.

which carries a pencil or point. The instrument is used as follows :—Let A C be the major, and H B the minor axis of an ellipse :
lay the cross of the trammel on these lines, so that the centre lines of
it may coincide with them; then adjust the sliders of the tracer, so
that the distance between *k* and *m* may be equal to half the major
axis, and the distance between *l* and *m* equal to half the minor

axis; then by moving the bar round, the pencil in the slider will describe the ellipse.

PROBLEM XXVI.—*An ellipse may also be described by means of a string.*

Let A B (Fig. 48) be the major axis, and D C the minor axis of the ellipse, and F G its two foci. Take a string E G F and pass it over the pins, and tie the ends. together, so that when doubled it may be equal to the distance from the focus F to the end of the axis, B; then putting a pencil in the bight or doubling of the string at H and carrying it round, the curve may be traced. This is based on the well known property of the ellipse, that the sum of any two lines drawn from the foci to any points in the circumference is the same.

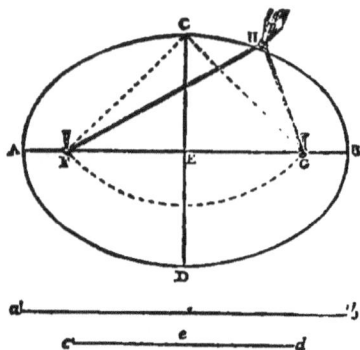

Fig. 48.

PROBLEM XXVII.—*The axes of an ellipse being given, to draw the curve by intersections.*

Let A C (Fig. 49) be the major axis, and D B half the minor axis. On the major axis construct the parallelogram A E F C, and make its height equal to D B. Divide A E and E B each into the same number of equal parts, and number the divisions from A and E respectively; then join A 1, 1 2, 2 3, etc., and their intersections will give points through which the curve may be drawn.

The points for a "raking" or rampant ellipse may also be found by the intersection of lines as shown at Fig. 50. Let A C be the major and E B the minor axis: draw A G and C H each parallel to B E, and equal to the semi-axis minor. Divide A D, the semi-axis major, and the lines A G and C H each into the same number of equal parts, in 1, 2, 3 and 4; then from E, through the divisions 1, 2, 3 and 4, on the semi-axis major A D, draw the lines E *h*, E *k*, E *l*, and E *m*; and from D, through the divisions 1, 2, 3 and 4 on the line A G, draw the lines 1, 2, 3 and 4 B; and the intersection of

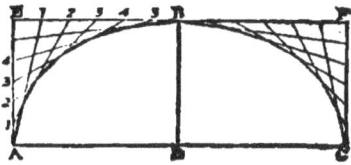 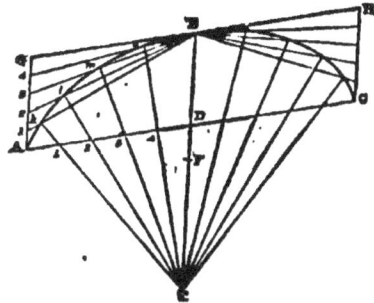

Fig. 49.   Fig. 50.

these with the lines E 1, 2, 3 and 4 in the points *h k l m*, will be points in the curve.

PROBLEM XXVIII.—*To describe with a compass a figure resembling the ellipse.*

Let A B (Fig. 51) be the given axis, which divide into three equal parts at the points *f g*. From these points as centres, with the radius *f* A, describe circles which intersect each other, and from the points of intersection through *f* and *g*, draw the diameters C *g* E, C*f* D. From C as a centre, with the radius C D, describe the arc D F, which

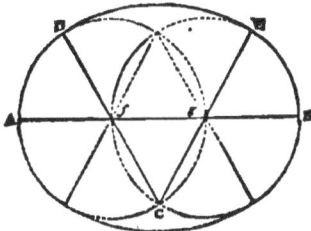 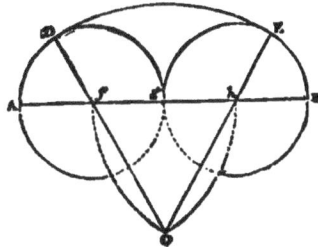

Fig. 51.   Fig. 52.

completes the semi-ellipse. The other half of the ellipse may be completed in the same manner, as shown by the dotted lines.

PROBLEM XXIX.—*Another method of describing a figure approaching the ellipse with a compass.*

The proportions of the ellipse may be varied by altering the ratio of the divisions of the diameter, as thus :—Divide the major axis of the ellipse A B (Fig. 52), into four equal parts, in the points *f g h*. On *f h* construct an equilateral triangle *f* c *h*, and produce

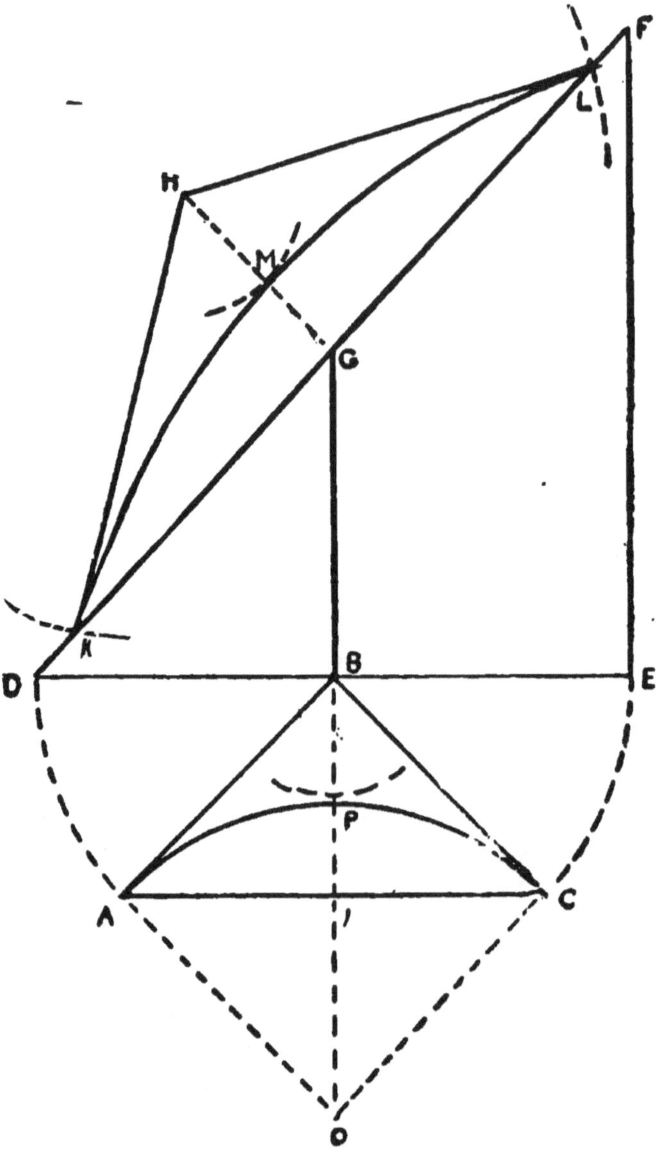

Fig. 53.

the sides of the triangle c ƒ, c *h* indefinitely, as to D and E. Then from the centres ƒ and *h*, with the radius A ƒ, describe the circles A D *g*, B E *g*; and from the centre C, with the radius C D, describe the arc D E to complete the semi-ellipse. The other half may be completed in the same manner. By this method of construction the minor axis is to the major axis as 14 to 22.

The following problems, which relate more particularly to hand-railing, should be thoroughly mastered before passing to actual work.

A tangent is a line touching a circle at right angles to the radius as shown at Fig. 55.

To construct Fig. 53: From the centre O with the radius O A, describe a quarter-circle A P C; draw tangents A B and C B; join A C; through the point B draw a straight line parallel to A C; with the centre B, with the radius B A, describe the arcs A D and C E; at the point E erect the perpendicular E F at right angles to D E to any desired height; in laying down a hand-rail this height will be the number of risers contained in the wreath; let F be the given height (this being one pitch); join F D; extend O B to G; draw G H at right angles to F D; make G H equal to B I; with the centre H and radius D G describe arcs cutting D F at K and L; draw H L and H K, which are the tangents on the pitch, and which when placed in position would stand plumb over A B C.

To construct Fig. 54, proceed as above until the height is located. It will be seen that here B G is lifted higher, making the pitch line and tangents F G and D G of unequal length. To obtain the angle, continue B G to H, making B H equal to E F; from H draw the line H J to any distance at right angles to D G. With the centre G and radius G F, describe an arc cutting the line H J at I; join I G and I D, and the angle is completed.

An easy way to prove these problems is to draw them on common thick paper; then take a knife, and cut out the angle D E F, place it perpendicularly over A B C, bringing D over A, and E over C; then cut out the angle H L K, and if drawn correctly it will lie on the pitch lines and fit the sides exactly.

To draw the curve line the quickest and most practical method

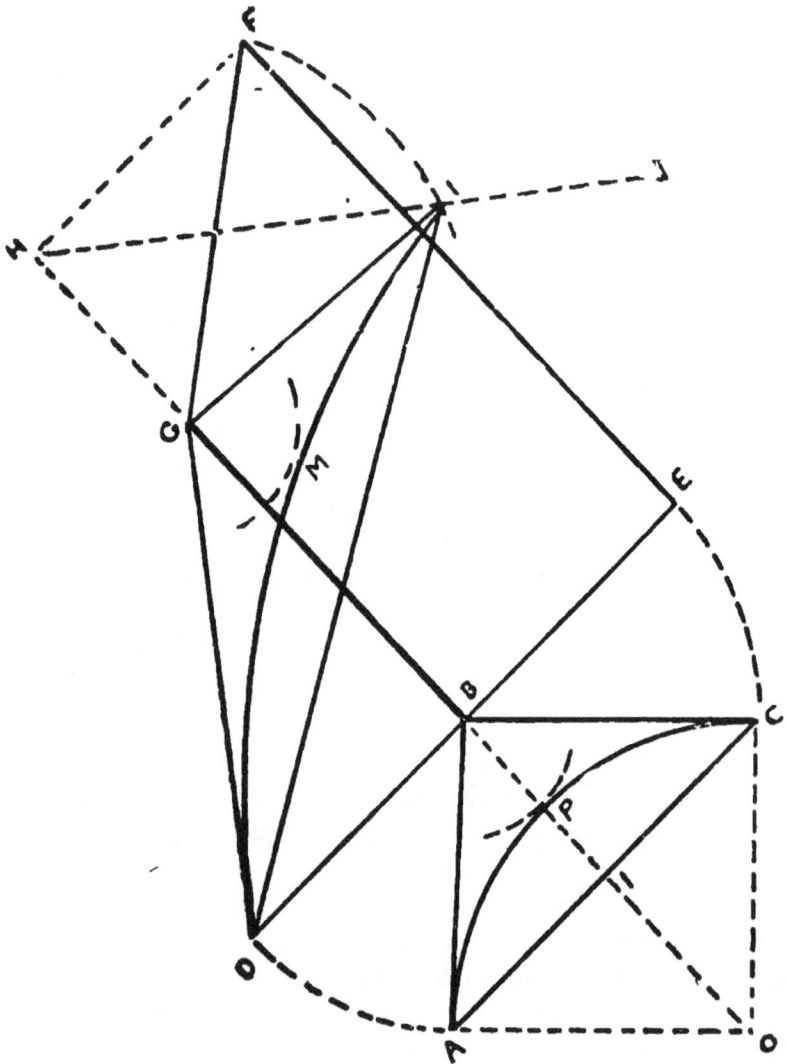

Fig. 54.

is to take B as a centre, and with a radius of B P, describe
an arc, touching the curve A P C, in the angle A B C, from II, as a
centre, with the same radius, describe an arc cutting H G at M;
then, take a thin flexible strip of wood of an even thickness, bend it
until it touches the points K I M; mark around it with a pencil,
and the curve is completed, and near enough to absolute accuracy
for all practical purposes.

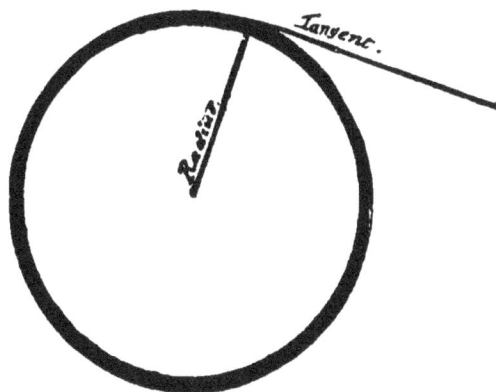

Fig. 55.

PART II.

TO MAKE THE WREATHS FOR STAIRS WITH FOUR WINDERS IN THE WELL.

FIGURE 1.—Strike the quarter circle *A B C*, which is the centre line of the rail, and also describe the inside and outside lines of rail *b b b*, and mark the risers at *A B* and *C*. Draw the lines *A D* and *C D*.

FIGURE 2.—Draw the straight line *B K* equal to the tread of one step, and set up *K A* equal to the rise of one step. Draw the line *A C*, and make *A D* and *D C* equal to the length *A D* and *D C* in Figure 1. Set up the height of two risers from *C* to *F*. Draw the line *A B*, and mark the point *G* at one third the distance from *A* to *B*, and draw the line *F G*. Continue *C A* to the point *H*, and from *D* draw the perpendicular line *D J*. Make the point at *H* square to *G H*, and ease off the rail from *I* to *H*, and draw the top and bottom side, by setting half the thickness of the rail each side the line *I H*.

FIGURE 1.—Draw the line *C I* through the centre *X*, and make *C I* equal to *D H* in Figure 2, and draw *H I*. Set up *C F* equal to *C F* in Figure 2, and *C J* equal to *D J* in Figure 2. Join *J I*, and from the point *F* draw the line *F G* parallel to *J I*, and join *G H*. Draw ordinate lines *a b b b* at any convenient distance apart all parallel to *G H*, one of which to pass from *D* to *X*. From the point *I* draw the line *I N*, square to *J I*, and from *G* with the distance *G H* cut the line *I, N* at *E*, and draw *G E*. Draw the perpendicular lines *a c* and from the points *c c c* draw ordinate lines *c d*, all parallel to *G'E*. Take the distance from *a* to *b b b*, and apply them from *c* to *d, d, d*, and draw the face mould

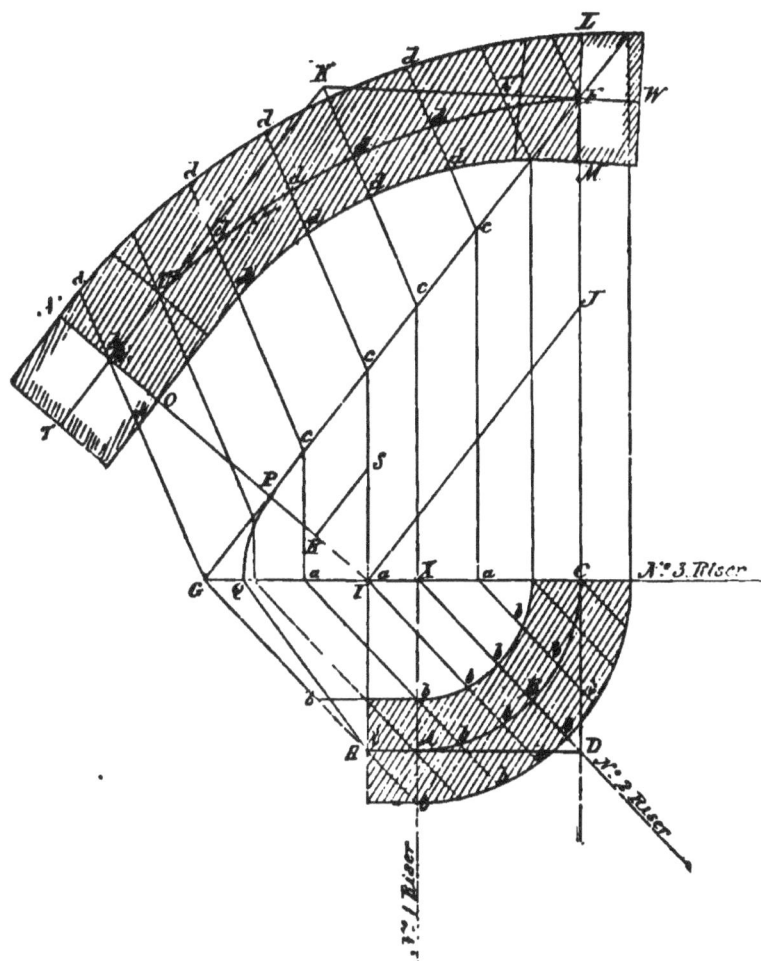

Fig. 1.

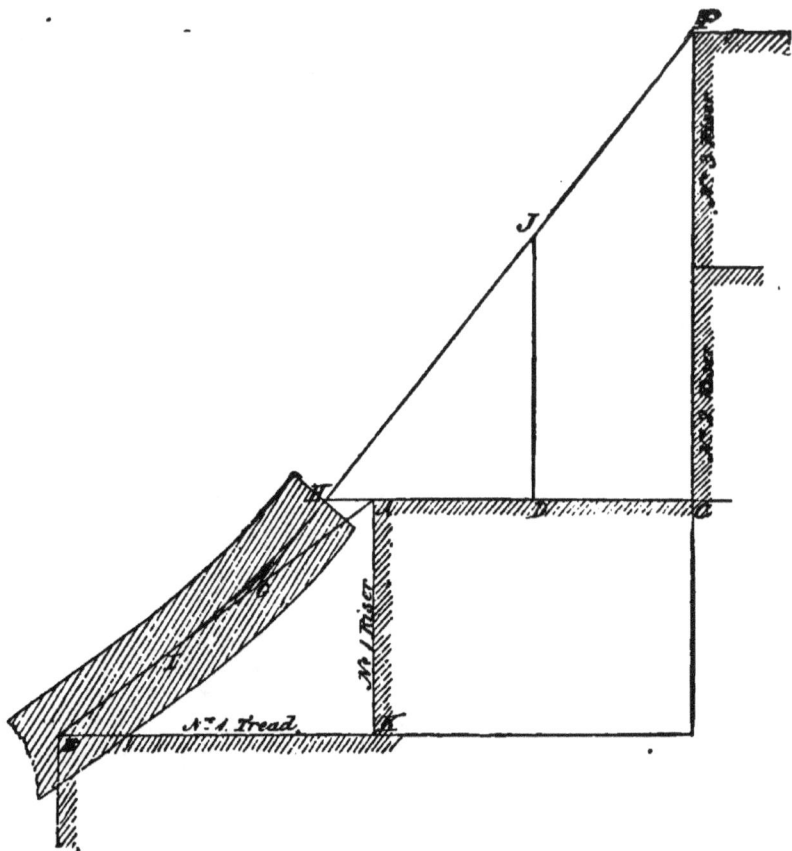

Fig. 2.

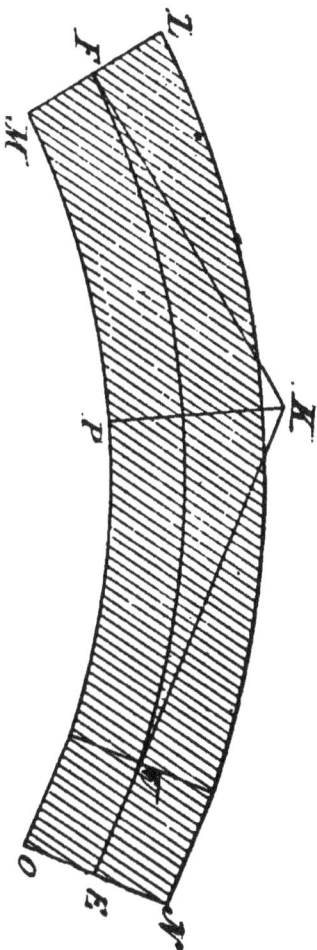

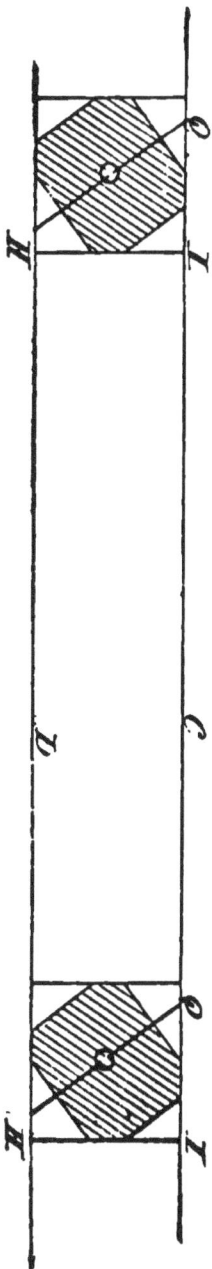

Fig. 3a.

Fig. 3.

through them. Take the distance *X D*, and mark it on the ordinate from *c* to *k*, and draw the lines *E K* and *K F*. Make the
joint *L M* square to *K F*, and the end *N O* square to *K E*.
From *I*, with the distance *I P*, describe the arc *P Q*, and draw
Q H, and apply this bevel *I Q H*, to both ends of the wreath
through the centre *O*, as shown at *I Q H* in Figure 4. Take the
distance *Q O* in Figure 4, and apply it from *I* to *R* in Figure 1;
draw *R S* parallel to *I J*, and cut it with the line *I, S*. Apply
the distance *R S* from *F* to *V*, and *F* to *W*; also from *E* to *T* and
E to *U*, and make the face mould the length from *T* to *W*.

FIGURE 3.—The centre line *E U F*, and the joints are the same
as *E U F* in Figure 1. Set the compasses to half the thickness of
of the plank *C D* in Figure 3*a*, and mark it each side the centre
line *E U F*, and make another face mould to this (Figure 3), mark
it on the plank, and cut it out square, and joint the ends square.
Now apply the face mould Figure 1 on the top side of wreath,
with the points *V* and *T* over the points *Q Q* at the ends of
wreath, and mark both edges; apply it on the underside with the
points *W* and *U* over the points *H H*, at the ends of wreath, and
mark both edges. Work the inside of the wreath with a round plane
to fit the rail on the plan, then work off the outside to the marks
on the top and bottom, and the bevels at the ends. To work
the top: At the mark *K P* on the face mould, take an equal portion off the top and bottom side of the wreath that will leave it the
thickness of the rail, and work it to the bevels marked on the ends,
gradually twisting from end to end. The bottom part can be
gauged from the top side.

TO MAKE THE WREATHS FOR STAIRS WITH SIX WINDERS IN THE WELL.

FIGURE 4.—Strike the quarter circle *A C*, which is the centre line of the rail, and also describe the inside and outside lines of rail, *b ὸ b*. Draw the lines *A D*, and *D C*.

FIGURE 5.—Draw the straight line *B K* equal to the tread of one step, and set up *K A* equal to the rise of one step. Draw the line *A C*, and make *A D* and *D C* equal to *A D* and *D C* in Figure 4; set up the height of 3 risers from *C* to *F*; draw the line *A B*, and mark the point *G* at one-third the distance from *A* to *B*, and draw the line *G F*; erect the perpendicular line *D J*. Joint the rail square at *H*, (which eases the rail better than if jointed at the spring of the well at *A*). Ease off the rail from *H* to *E*, and draw *H I* parallel to *A C*.

FIGURE 4.—Draw the line *C G* through the centre *X*, and make *C I* equal to *L H* in Figure 2, and draw *I H*. Set up *C F* equal to *I F* in Figure 5, and *C J* equal to *L J*; join *J I*, and from the point *F* draw the line *F G* parallel to *J I*, and join *G H*.

Draw ordinate lines *a b b b*, at any convenient distance apart, all parallel to *G H*, one of which to pass from *D* to *X*. From the point *I* draw the line *I N* square to *I J*, and from *G*, with the distance *G H*, cut the line *I N* at *E*, and draw *G E*.

Draw the perpendicular lines *a c*, and from the points *c c c*, draw ordinate line *c* to *d d d*, all parallel to *G E*. Take the distances from *a* to *b b b*, and apply them from *c* to *d d d*, and draw the face mould through them.

Take the distance *X D*, and mark it on the ordinate from *c* to *K*, and draw the line *E K* and *K F*. Make the joint *L M* square to *K F*, and the end *N O* square to *K E*.

From *I*, with the distance *I P*, describe the arc *P Q*, and draw *Q H*. Apply the bevel *I Q H* to both ends the wreath through the

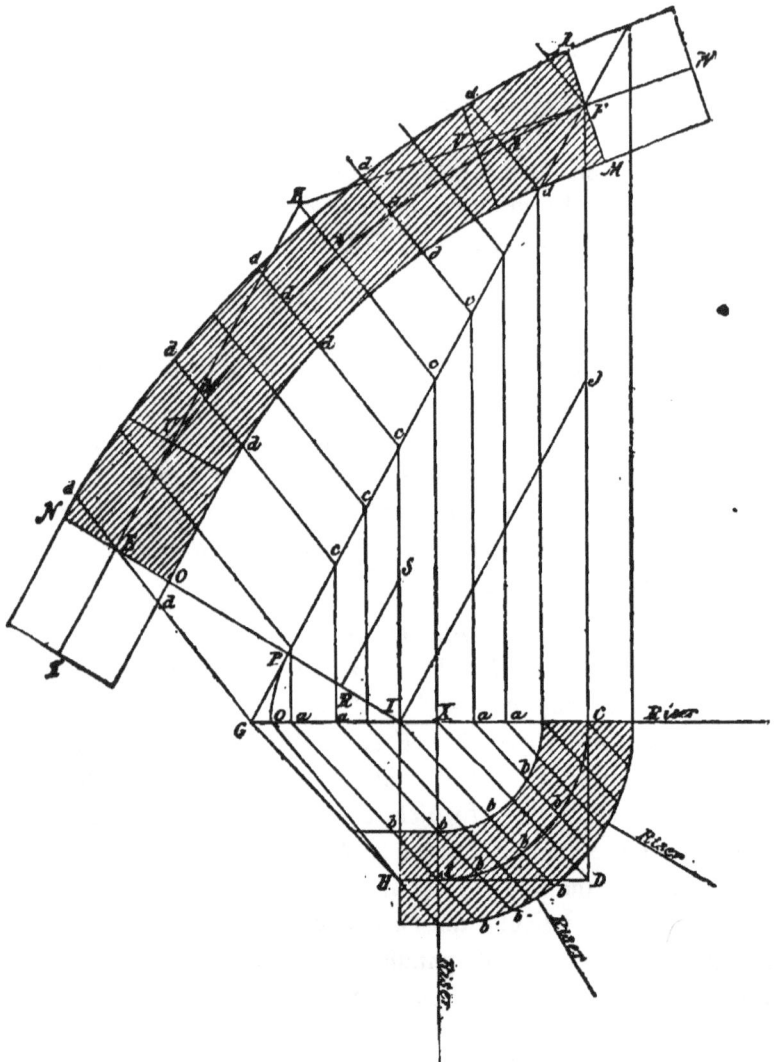

Fig. 4.

Fig. 6.

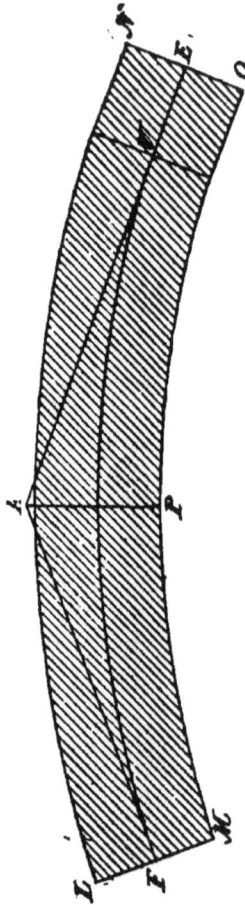

Fig. 6.

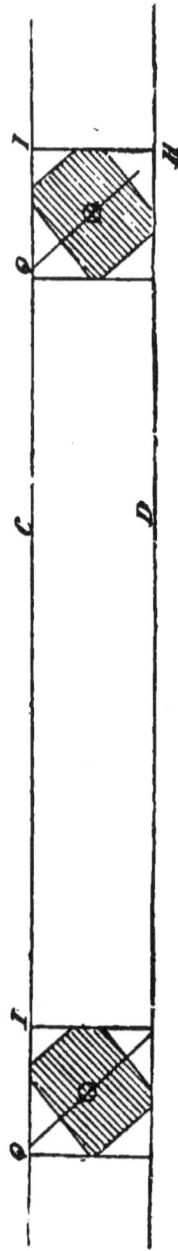

Fig. 7.

centre *O*, as shown at *I Q H* in Figure 4. Take the distance *Q O* in Figure 4, and apply it from *I R* in Figure 1, and draw *R S* parallel to *G F*, and cut it with the line *I S*. Apply the distance *R S* from *F* to *V*, and *F* to *W*, also from *E* to *T* and *E* to *U*, and make the face mould from *T* to *W*.

FIGURE 6.—The centre line *E U F* and the joints at the ends, are the same as *E U F* in Figure 1. Set the compasses to half the thickness of the plank *C D* in Figure 7, and mark it each side the centre line *E U F*. Make another face mould, (to Figure 6) mark it on the plank, cut it out square, and joint the ends square.

Apply the face mould, Figure 4, on the top side the wreath, with the points *T* and *V* over the points *Q Q* at the ends of the wreath, and mark both edges. Apply it on the underside the wreath, with the points *W* and *U* over the points *H H* at the ends of the wreath, and mark it.

Work the inside the wreath with a round plane to fit the inside curve on plan, then work off the outside to the marks on the top and bottom and the bevels at the ends.

To work the top, at the mark *K P* on the face mould, take an equal portion off the top and bottom side the wreath, that will leave it the thickness of the rail, and work it from the bevels at each end, through *K P*, gradually twisting from end to end. The bottom part can then be gauged from the top side.

TO MAKE THE WREATHS FOR STAIRS WITH THREE WINDERS IN HALF THE WELL, AND LANDING IN THE CENTRE OF THE WELL.

(THIS IS ONE OF THE MOST DIFFICULT RAILS THAT CAN BE MADE.)

FIGURE 8.—Strike out the semi-circle A B C, which is the centre of rail, and also describe the inside and outside lines of rail. Draw the lines C E, A D, and D E.

FIGURE 9 (*See Folding Plate*).—Draw the straight line O K equal to the tread of one step, and set up $K A$ equal to the rise of one step. Draw the line A E, and make $A D$—$D B$ and $B E$ equal to $A D$—$D B$ and $B E$ in Figure 8. Set up the height of 3 risers from E to M, and six inches more from M to C, (this makes the rail 3 feet high on the landing, or six inches higher than the rail is on the stairs, which should be 2½ feet plumb with the face of the risers.)

Draw the line A O, and mark the point G about one-third the distance from A to O, and join C G. Make the joint at H which is a little above the spring of the well at A, and ease off the rail as shown at G.

Draw the line $H L I$ parallel to $A B$, and erect the perpendicular lines $L J$ and $I F$. Now from the base line $H L I$ and the height $L J$ and $I F$ get the face mould and work the bottom wreath, as shown and described in Figure 5, from the base line $H L I$ and the heights $L J$ and $I F$.

To work the top wreath proceed as follows, viz.:—

FIGURE 10.—Lay down the plan of the rail, the centre line B C F, being the same as $B C F$, in Figure 8, and make the straight part about 5 inches from C to F. Draw the lines $F A$-A B-B E and E F. Take the height $N C$ in Figure 9, and apply it from F to D, and join $A D$. Draw $D G$ square to $A D$, and make $D G$ equal to $F E$.

Fig. 10.

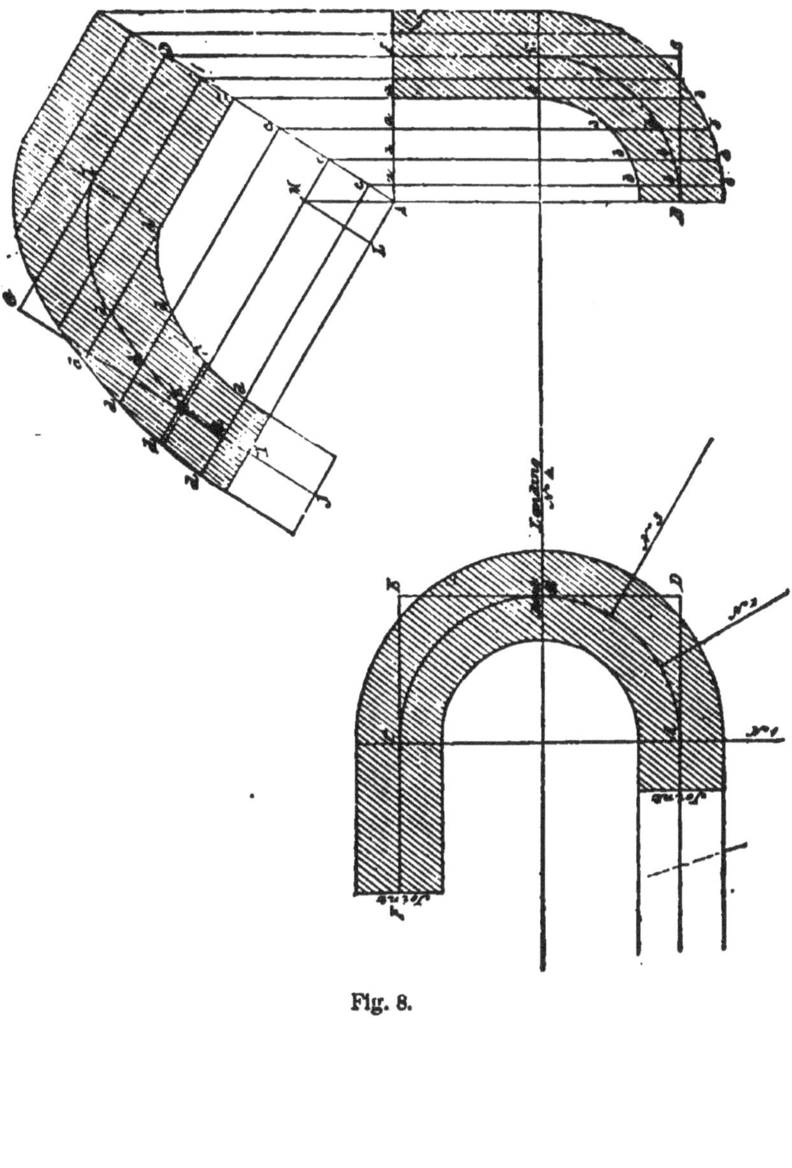

Fig. 8.

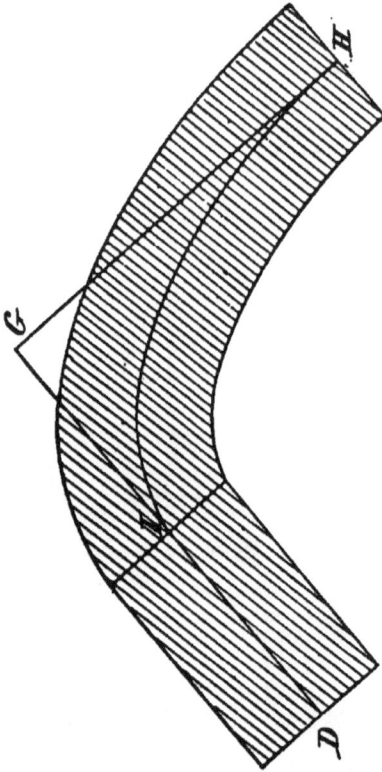

Fig. 12.

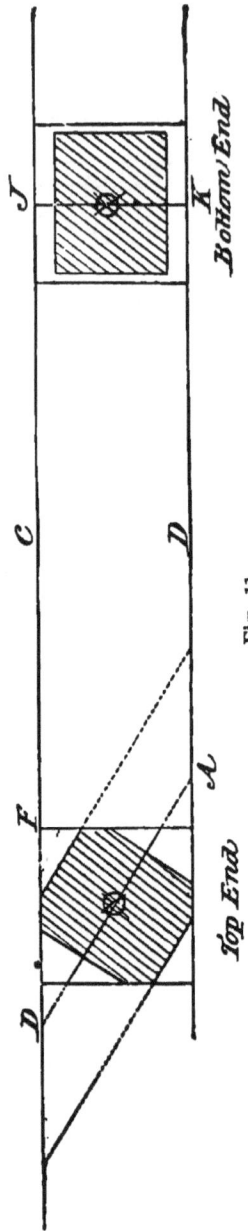

Fig. 11.

Draw ordinate lines *a b b b* parallel to *A B*, and draw the lines *a c*, and from *c c*, draw ordinate *c d d d*, parallel to *G D*. Apply the distance from *a* to *b, b, b*, on to the ordinates from *c* to *d, d, d*, and draw the face mould through them. Make the joints at *H* and *D* square, as shown.

FIGURE 11.—Shows the ends of the wreath. Take the bevel *F D A* in Figure 10 and apply it at the top end of the wreath through the centre *O* at *F D A*, and mark the size of the rail on as shown.

FIGURE 12.—Is the face mould to cut the wreath out square; the centre line *D I H*, and the joints are the same as the centre line *D I H* in Figure 3; put half the thickness of the plank each side the centre line *D I H*, and cut out the wreath, Figure 8: Take half the thickness of the plank from *A* to *L*, draw *L M* parallel to *A D*, and draw *A M*, apply the distance *L M* to *H J*, and *H K*, and make the face mould from *J* to *D*.

Apply Figure 10 face mould on the top side of wreath, with the points *D* and *J*, over the points *D J* in Figure 11, and mark it. Apply it on the underside with the points *D* and *K* over the points *A* and *K* in Figure 11.

Work it as described in the preceding Figures 6 and 7,. to the bevels at the ends as shown, and work the part from *D* to *I* straight; then work the top gradually twisting from *I* to H in Figure 12.

TO MAKE THE WREATHS FOR STAIRS WITH ONE RISER IN THE CENTRE OF WELL.

Figure 13 (*See Folding Plate*).—Mark the plan of the rail as *A B C D*, and *E F G* the centre line, the straight part *E F* being the width of one step. Continue *E F* to *H* and draw *G H*. Draw *D I* parallel to *E F*, and draw the lines *A J* and *F O*. Make *O K* equal to the rise of one step, and draw *J K*, set up *G L* equal to the rise of two steps. From the point *L* make the line *L I* parallel to *J K* and join *I E*. Draw ordinate lines as 1, 2, 3, 4, at any convenient distance apart all parallel to *I E*, draw the perpendicular lines 1, 5, etc., to intersect the line *I L*. From the point *J*, draw the line *J M*, square to *J K*, and from *I*, with the distance *I E*, cut *J M* at the point *N*, and draw *I N*; draw ordinate lines 5, 6, 7, 8, etc., all parallel to *I N*. Take the distances 1, 2, 3, 4, and apply them from 5 to 6, 7, 8, etc., and draw the inside and outside and centre line of face mould through them.

Take the length of the ordinate *O H*, and apply it on the ordinate from *P* to *Q*, and draw the lines *N Q*, and *L Q*. Make the joint *R S* square to *Q L*, and the joint *M N* square to *Q N*.

Figure 14.—The centre line *N* 7 7 7 *L*, and the joints at ends, are the same as Figure 13.

Set the compasses to half the thickness of the plank, (as shown at *C D* in Figure 15), and apply it on each side the centre line *L* 7 7 7 *N*, and draw the inside and outside lines. Mark this face mould on the plank, and cut it out square, and joint the ends square.

To put the twist on the wreath, proceed as follows:—

In Figure 13 from the centre *J*, describe the arc *T U*, and draw *U E*. Continue the lines *I E*, and *G H*, until they meet at *W*. Take the length *G L* and apply it from *G* to *V*, and draw *V W*. From *G* draw *G X* square to *V W*, and describe the arc *X Y*, and draw *Y I*.

Fig. 14.

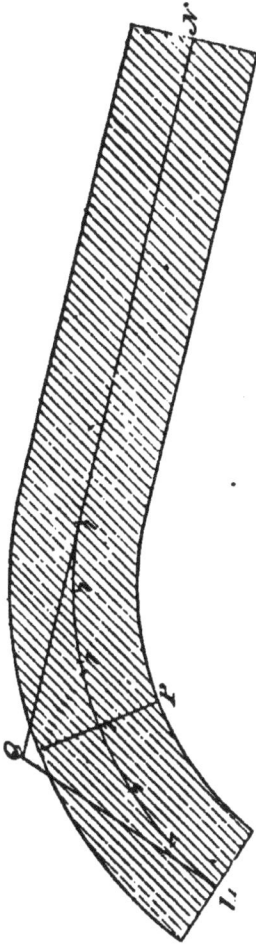

Fig. 15.

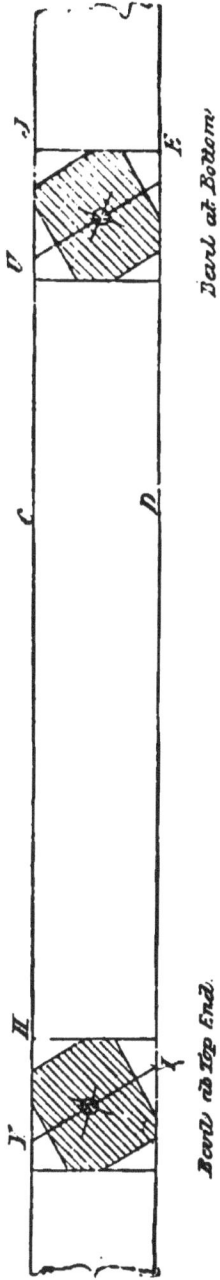

Take the bevel $U\ J\ E$, and apply it on the bottom end of the wreath in Figure 15, at $U\ J\ E$, through the centre O, and mark half the width of the rail on each side $U\ E$, also mark the top and bottom line of rail square to $U\ E$. Apply the bevel $H\ Y\ I$ in Figure 1 to $H\ Y\ I$, at the top end of the wreath in Figure 15, through the centre O, and mark the size of the rail on as shown. Mark the line $Q\ P$ on both sides the wreath. Take the distance $O\ U$ in Figure 15, and apply it from J to a in Figure 13, draw $a\ b$ parallel to $J\ K$, and draw $J\ b$, apply the distance $a\ b$ from N to c and N to d. Apply the distance $O\ Y$, in Figure 15, to $X\ e$ in Figure 13, and draw $e\ f$ parallel to $G\ V$, and apply the distance $X\ f$ to $L\ g$, and $L\ h$. Make a face mould to Figure 13, the full length from c to h, and place it on the top of the wreath, with the point g over the point Y, in Figure 15, at the top end of the wreath, and the point c over the point U, at the bottom end and mark both edges, then mark it on the underside the wreath, with the point h over the point I in Figure 15, and the point c over the point E.

First work off the inside of the wreath with a round plane that will fit the rail on the plan, then work off the outside part. To work the top at the mark $Q\ P$, take an equal portion off the top and bottom side of the wreath that will leave it the finished thickness of the rail, then work from the bevels marked at each end, through the part taken off at $Q\ P$ and the bottom part can be gauged from the top.

TO MAKE THE WREATHS FOR LEVEL LANDING STAIRS. THE DISTANCE BETWEEN CENTRE OF RAILS ACROSS THE WELL BEING EQUAL TO THE TREAD OF A STEP, OR NEARLY SO.

FIGURE 16.—$A B C$ is the centre of the rail, $A B$ being equal to the tread of one step. Strike out the inside and outside of the rail b $b b b$. Draw $C D$ and $D A$. Set up $E F$ equal to one riser, and draw $D F$ to G, and draw the line $C G$. Draw ordinate lines $a b b$, at any convenient distance apart, and draw the lines a, c. From the points $c c$ draw ordinate lines $c d$, square to the line $D G$. Take the distance $a b b$, and apply them on the ordinates from c to $d d$, and draw the face mould through them.

Take the thickness of the plank $C A$ in Figure 18, and apply it from D to I in Figure 16, draw $I H$ parallel to $D F$, and cut it with the line of $D H$. Take half of $I H$ and apply it from K to L, and K to J. Make the face mould the length from J to G, and mark it on the top side of the plank, and cut the two joints at J and G square.

Take the bevel $D J E$ in Figure 16, and apply it to the top joint at $D F E$, as shown in Figure 18. The bottom end will be square, as shown in Figure 17.

Apply the face mould underside the wreath with the point L at the lower end of wreath. Cut it out bevel (as this kind of wreath is less labor cut bevel than having to bevel it if cut out square).

Take an equal portion off the top and bottom side of the shank part of wreath to leave the rail the finished thickness as shown in Figure 17, at the bottom end, then gradually work the top side twisted to the bevel at the top end, so that when the wreath is put up to its proper pitch the top side is level across. Gauge the bottom from the top side.

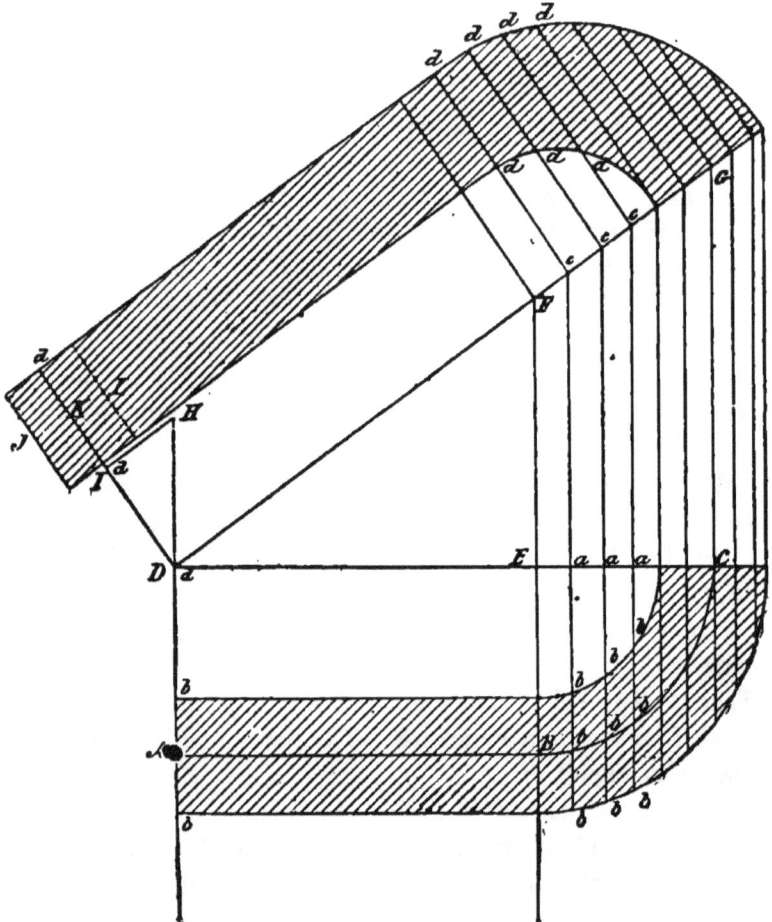

Fig. 16.

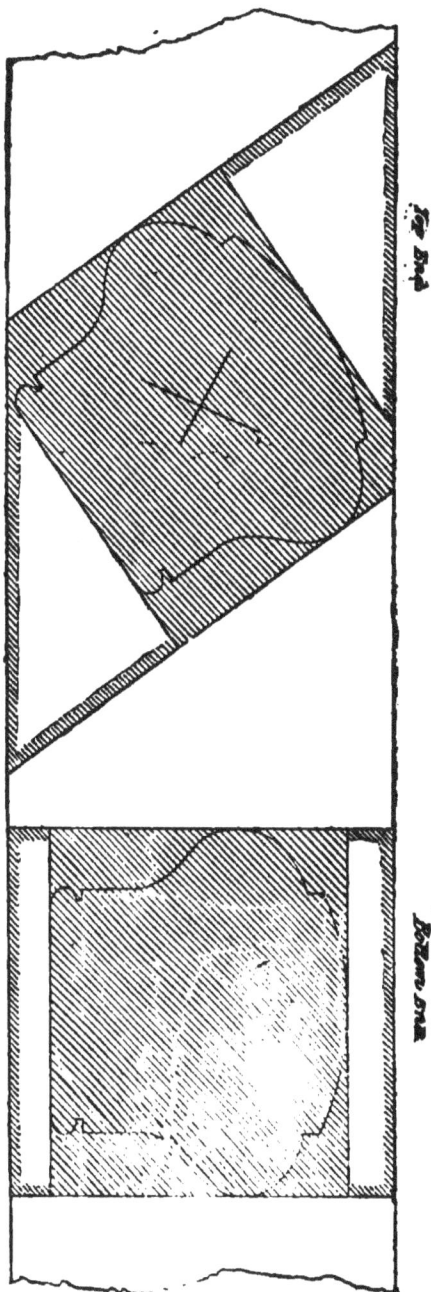

Fig. 17.

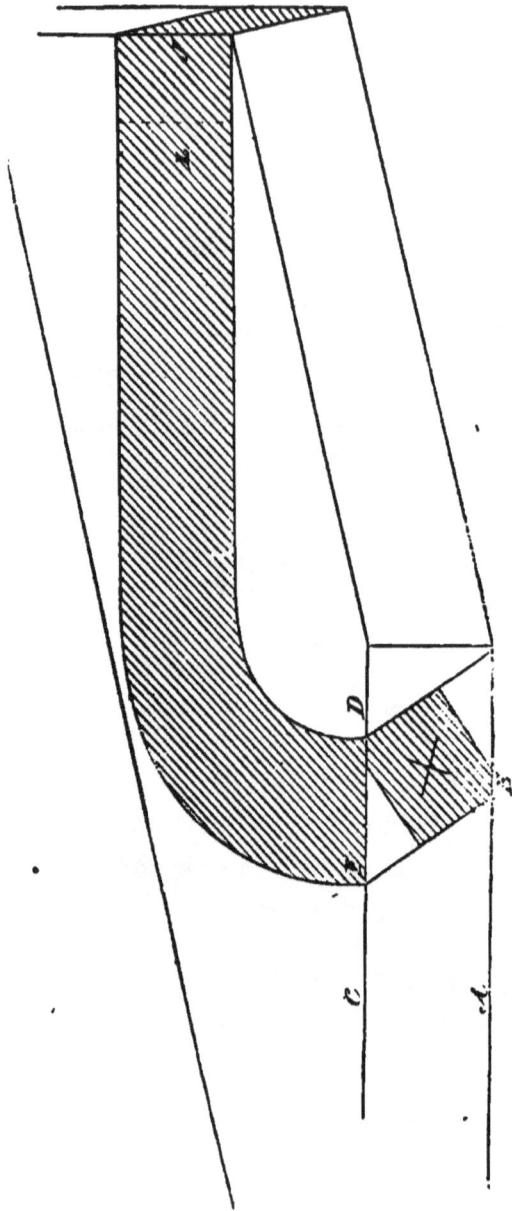

NOTE.—The distance across the well between centre of rails should not exceed the tread of a step for level landing stairs, or the wreath has the appearance of dropping.

If the distance is less than the tread of a step bevel ordinate lines would be required, after the manner shown in Figures 13, 14, and 15.

TO MAKE THE WREATH FOR QUARTER-LANDING STAIRS, HAVING ONE WINDER IN THE WELL.

FIGURE 19.—From the centre O strike the quarter circle A B, the centre of the rail, and draw the lines O A—O B—B C and A C, and draw the inside and outside curve of rail.

FIGURE 20 (*See Folding Plate*).—Make D E equal to the tread of one step, and E A equal to one riser.

Draw the line A B and make A B and B C equal to A B and B C in Figure 19.

Set up the height of 2 risers from B to F, and draw the line F G, to pitch of the stairs. Join A D and mark the point H about one-third the distance from A to D, then mark the point I about one-third the distance from F to G, and draw the line I H.

Draw the perpendicular line C J, and continue A B to K.

Joint the end at K square to J K and ease off the rail at H.

Joint the top length at the point L and ease off the rail at I.

FIGURE 19.—Make B K equal to C K in Figure 20. Make B J equal to C J in Figure 20, and B M equal to B M in Figure 20. Draw the line J K, and from the point M draw M D parallel to J K. Draw K F square to J K, and from the point D with the distance D E cut it at the point G. Draw the ordinate a b b b parallel to D E, draw the lines a c, and from c draw ordinates c d d d parallel to D G.

Apply the distances a b b b, on to the ordinates from c to d d d, and draw the face mould through the points d d d.

Mark the length of the ordinate O C, on to the ordinate from c to H, and draw H G and H M. Joint the end at G square to G H.

Take the distance M L in Figure 20, and apply it from M to T, and joint the end at T square to H T. Apply the bevel K I E to both ends the wreath as shown in Figure 22.

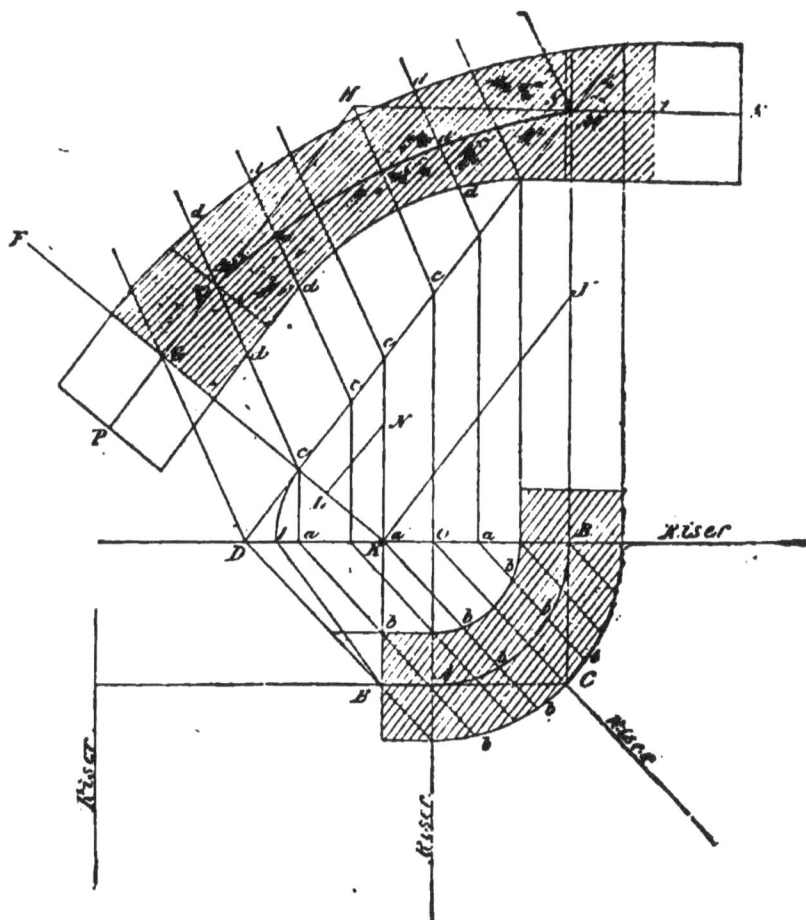

Fig. 19.

The face mould Figure 21, is got from the centre line of Figure 19, as before described, and the joints at T and G are the same as in Figure 19.

Apply the distance $O\,I$, in Figure 22 to $K\,L$ in Figure 19, draw $L\,N$, and mark the length $L\,N$ from G to P and G to Q, also from T to S and T to R. Make the face mould Figure 19, the length from P to R.

Fig. 21.

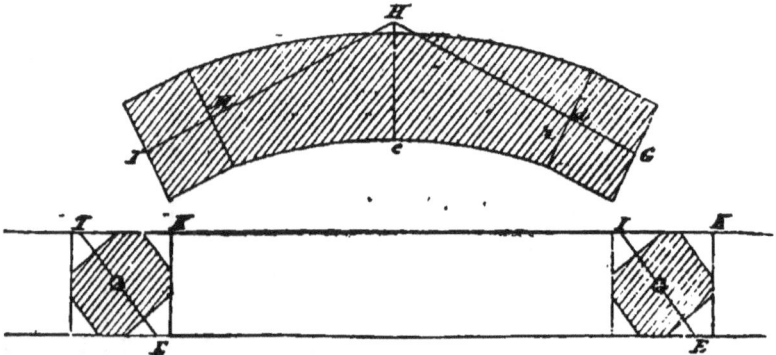

Fig. 22.

Cut the wreath out square to Figure 21, face mould and joint the ends. Apply Figure 19 face mould on the top side of wreath with the points $S\,P$, over $I\,I$ at the ends of wreath, then apply it underside with the points $R\,Q$, over $E\,E$ at the ends.

Work the inside the wreath first with a round plan to fit the rail on plan, and work the top side straight from T to M and d to G in Figure 21, and gradually work it twisted to the bevels at each end, and take an equal portion off the top and bottom side at $c\,H$ to leave the rail the finished thickness.

TO MAKE THE WREATH FOR QUARTER LAND-ING STAIRS, WITH THE RISERS AS SHOWN IN FIGURE 23.

FIGURE 23.—From the centre O strike the quarter circle A B, which is the centre of the rail, draw the lines O A—O B—B C and A C, and draw the straight part each end about 3 inches long from A to D, and B to E, then draw the inside and outside lines of rail.

FIGURE 24 (*See Folding Plate*)—Make D E equal to one tread, and E A equal to one riser. Draw the line A B, the parts A C and C B, being equal to A C and C B in Figure 1. Set up B F equal to one riser and from F draw the line G F H, to the pitch of the stairs, and draw C G.

Join A D, and mark the point I about half way between A and D, and draw I G.

Make K J equal to A D in Figure 23, and joint the end at J, square to I G, draw the line J N.

Make F L equal to E B in Figure 23, draw L M and joint the end at M square to F H.

FIGURE 23.—Draw the line D J, set up B F equal to O G in Figure 24, and joint J F.

Set up B G equal to N F in Figure 24, and from the point G draw G H parallel to J F.

From J, draw J I square to J F, and from H with the distance H D, cut J I at K, and draw H K.

Draw the ordinates a b b b, parallel to H D, draw the lines a c, and from the point c, draw ordinate c d d d parallel to H K; apply the distance a b b b, on to the ordinates c d d d, and draw the face mould through the points d d d.

Mark the length of the ordinate a C, on the ordinate c L, and draw L K and L G.

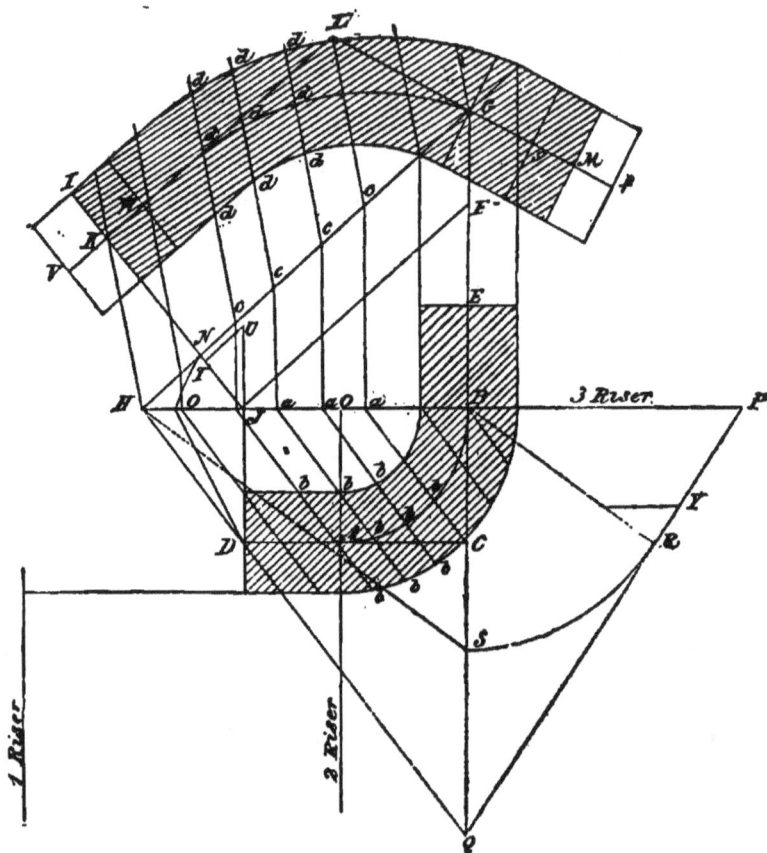

Fig. 23.

Make G M equal to F' M in Figure 24. Joint the end at K square to K L, and the end at M square to L M.

Get the level O J D for the bottom end, and B S O for the top end of wreath, as described in Figure 15, and apply them at the ends as shown in Figure 26.

The face mould Figure 25, is obtained from the centre line of Figure 23, as described in previous examples.

Fig. 25.

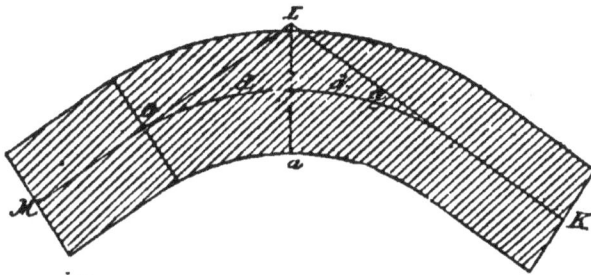

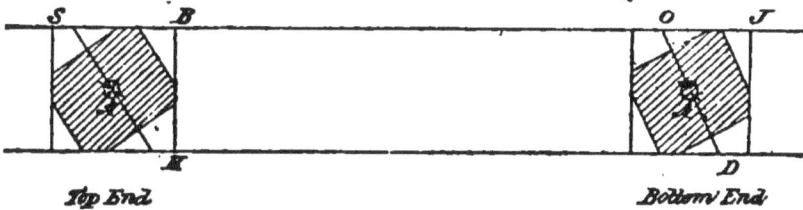

Fig. 26.

Apply the distance T U from K to V, and K to W, and the length R Y apply from M to g, and M to f, and make Figure 23, face mould the length from V to f.

Cut the wreath out square, to Figure 25, face mould, and joint the ends.

Apply Figure 23, face mould on the top of wreath, with the points f and W over the points S and O at the ends of wreath, then mark it underside with the points g and V, over the point H and D at the ends of wreath. Proceed to work the wreath as before described, to the bevels at the ends, and take as much of the top and bottom sides of the wreath at L a as will leave the rail the finished thickness.

www.ingramcontent.com/pod-product-compliance
Lightning Source LLC
Chambersburg PA
CBHW022009190326
41519CB00010B/1449